DeltaScience ContentReaders

Properties of Matter

Contents

Preview the Book 2
What Is Matter? 3
   Physical Properties of Matter 4
      Mass 5
      Volume 6
      Density 7
      State of Matter..................... 9
      Other Physical Properties 12
   Chemical Properties of Matter 13

Compare and Contrast 14
What Is Matter Made Of? 15
   Atoms 16
   Elements 17
   Grouping the Elements 18
   Molecules 21
   Compounds 22

Glossary 24

Build Reading Skills
Preview the Book

You read nonfiction books like this one to learn new information. Previewing the book will prepare you to understand and remember what you read.

First, look at the title, front cover, and table of contents. Based on these features, what topics do you predict you will read about? Think about what you already know about matter and its properties.

Next, look through the book page by page. Skim the headings and the words in bold type. Glance at the pictures and read some of the captions. Notice that each main section of the book begins with a two-page photograph. What other special features do you see in the book?

Headings, captions, and other features of nonfiction books are like signposts on a journey. They can help you find your way through new information. They can also help you connect new information with what you already know.

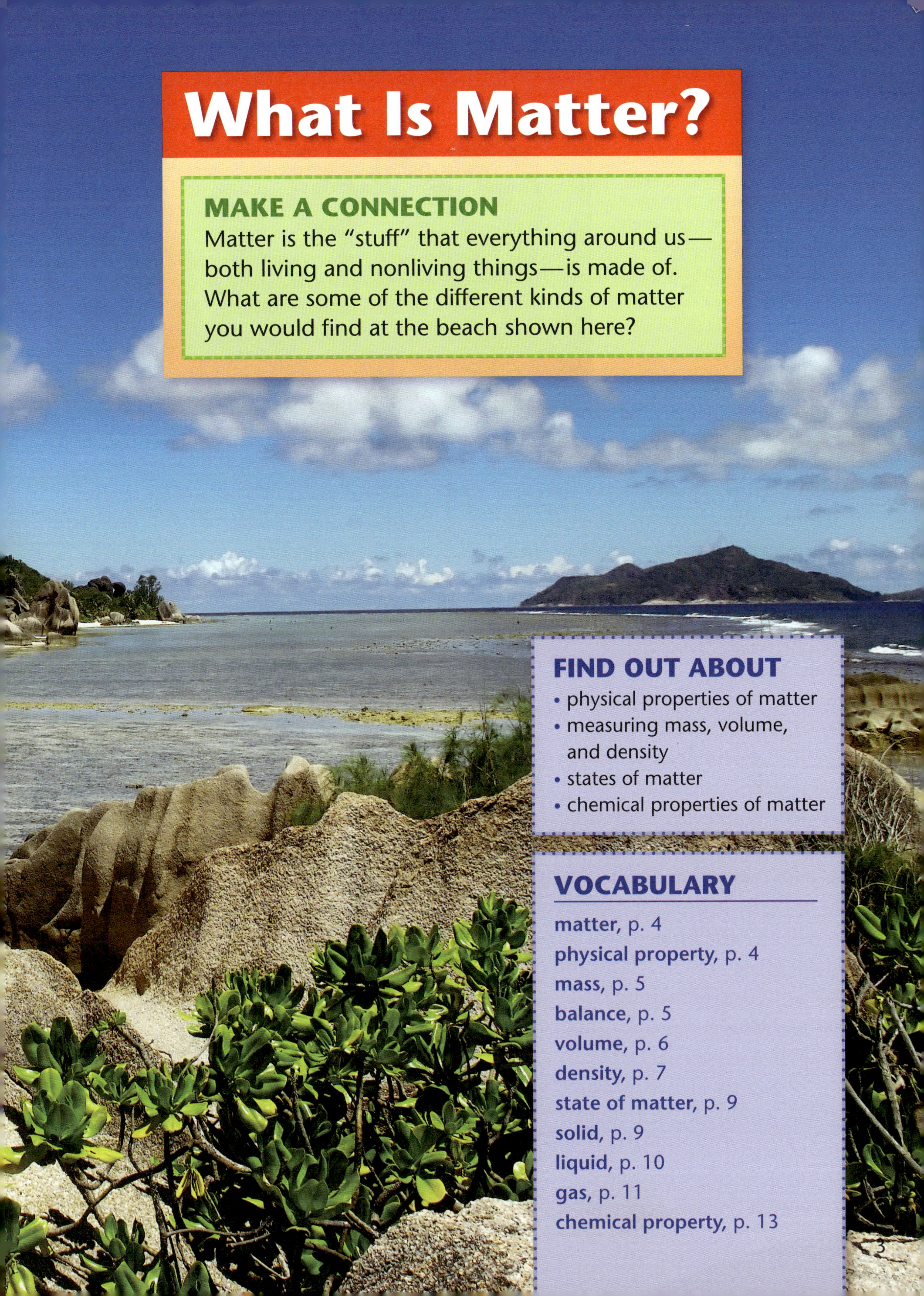

What Is Matter?

MAKE A CONNECTION
Matter is the "stuff" that everything around us—both living and nonliving things—is made of. What are some of the different kinds of matter you would find at the beach shown here?

FIND OUT ABOUT
- physical properties of matter
- measuring mass, volume, and density
- states of matter
- chemical properties of matter

VOCABULARY

matter, p. 4
physical property, p. 4
mass, p. 5
balance, p. 5
volume, p. 6
density, p. 7
state of matter, p. 9
solid, p. 9
liquid, p. 10
gas, p. 11
chemical property, p. 13

Each object in the aquarium is made of matter that we can describe by its physical properties. The pebbles are round, the plant is tall, the fish are orange, and the glass is smooth. ▶

Physical Properties of Matter

What do fish, pebbles, water, and gas bubbles have in common? They are all made of matter. **Matter** makes up all the objects, materials, and living things found on Earth and out in space. Scientists define matter as anything that has mass and takes up space.

We can identify and describe different types of matter by their physical properties. A **physical property** is a characteristic of an object that can be observed with our senses or measured with tools. Color and shape are physical properties that we can observe with our sense of sight. Texture, or how an object feels, is a physical property we can observe with our sense of touch. Taste and odor are also physical properties.

Size and temperature are examples of physical properties that can be measured using tools. Size, such as an object's length or width, can be measured with a ruler. Temperature, or how hot or cold something is, can be measured with a thermometer.

What kinds of things are *not* matter? Forms of energy, such as light, heat, and sound, are not matter. Forces, such as gravity and magnetism, are not matter either. Energy and forces do not have mass, do not take up space, and do not have physical properties.

 Define *physical property* and describe two physical properties of a pencil.

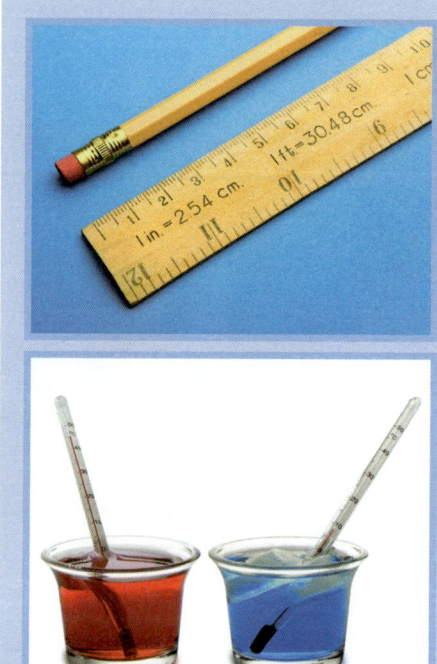

Tools help us measure the physical properties of matter. A ruler measures length, and a thermometer measures temperature.

4

Mass

All matter has mass. **Mass** is the amount of matter in an object. The more mass an object has, the heavier it feels. A large rock feels heavier than a feather does because the rock has a greater mass.

In the metric system, mass is measured in units called grams (g) and kilograms (kg). One kilogram equals 1,000 grams. One way to measure the mass of an object is to use a **balance**, a type of scale that compares the mass of two or more objects.

Suppose you want to use a balance to measure the mass of an apple. First, set the apple on one pan of the balance, causing that pan to tip down. Then place weights with known masses on the other pan, causing the pan with the apple to begin to rise. Keep adding weights one at a time until the two pans balance. The total mass of the weights equals the mass of the apple. An average apple has a mass of about 150 grams.

Mass is not the same as weight. Weight is a force—the force of gravity pulling on an object. However, weight is related to mass. The more mass an object has, the more it weighs.

 What is mass and how is it measured?

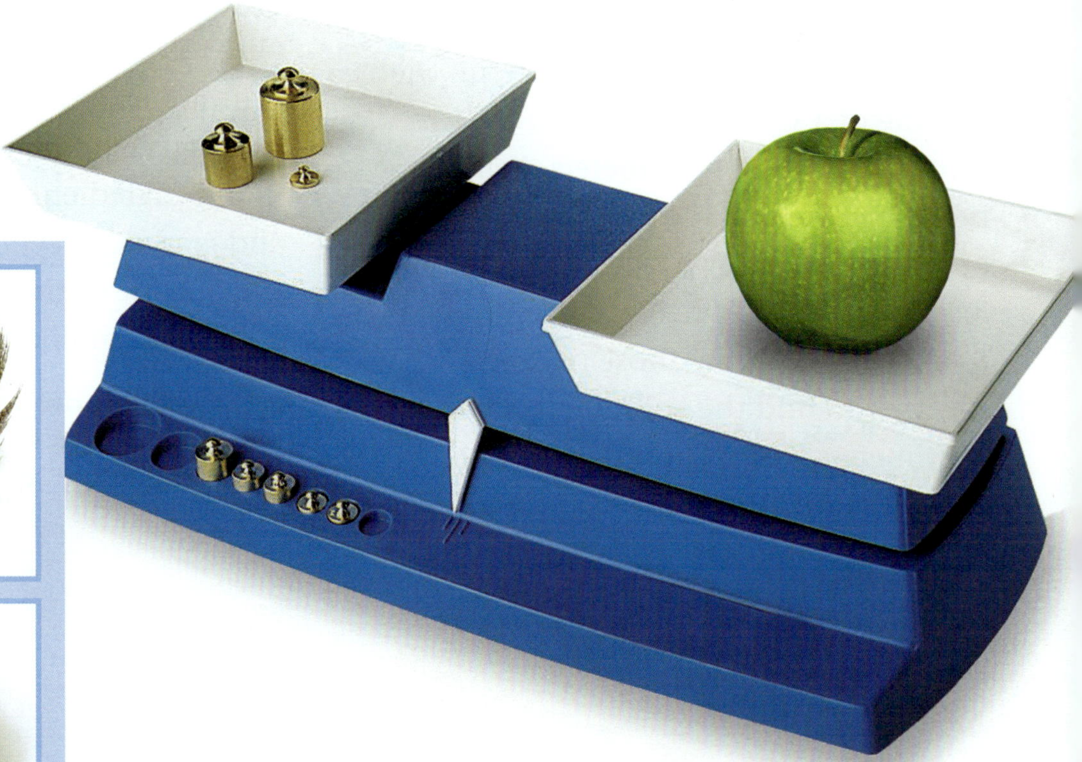

▲ A balance is a tool for measuring mass. Brass weights of known mass are placed in one pan to determine the mass of an object in the other pan.

A feather and a rock may be similar in size. However, the rock has more mass than the feather, so it feels heavier.

5

◀ Which sphere takes up the most space, the marble, the softball, or the beach ball? Sometimes you can see differences in volume without measuring.

Volume

In addition to having mass, all matter takes up space. The amount of space that matter takes up is called **volume**. Volume is another physical property.

Volume is measured in different ways, depending on the other properties of a material or object, such as its physical form or shape. In the metric system, the volume of a solid object is measured in units called cubic centimeters. You can figure out the volume of a block-shaped object using a ruler. Measure the object's length, width, and height in centimeters (cm). Then multiply those numbers together to find the volume in cubic centimeters (cm^3). Suppose you have a sponge that is 10 centimeters long, 7 centimeters wide, and 2 centimeters high. The volume of the sponge is

$$10 \text{ cm} \times 7 \text{ cm} \times 2 \text{ cm} = 140 \text{ cm}^3$$

The volume of a liquid is measured in units called milliliters (ml) and liters (L). One milliliter is equal to one cubic centimeter. One liter is 1,000 milliliters. Cooks use measuring cups to measure liquid ingredients. Scientists use beakers and graduated cylinders. The numbers on the outside of the container indicate the volume in milliliters or liters.

✓ **Explain how you could calculate the volume of a tissue box using a ruler.**

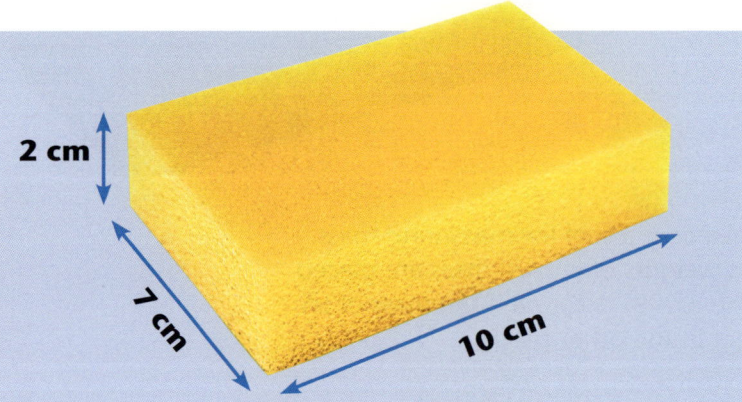

Volume is the amount of space an object takes up. To find the volume of this sponge, multiply the length times the width times the height.

▲ The volume of liquids can be found with tools such as measuring cups or graduated cylinders.

6

◀ Density is how heavy an object is for its size. A brick has more mass and weighs more than a wooden block with the same volume. So the density of brick is greater than the density of wood.

Density

Suppose you have a wooden block and a brick of the same size. Their volumes are the same, but the brick feels heavier. Why? The two objects take up equal amounts of space, but the brick has more matter, or mass, packed into that space. **Density** is a physical property that describes the amount of mass something has compared to its volume. Brick has a greater density than wood. Every material has its own density.

You can determine density by dividing the mass of an object by its volume. Imagine a solid cube made of the metal aluminum. Its mass is about 27 grams (g) and its volume is 9 cubic centimeters (cm^3). If you divide its mass by its volume— 27 g divided by 9 cm^3—you get 3 grams per cubic centimeter (g/cm^3). So the density of aluminum is about 3 g/cm^3.

The density of a material does not depend on the amount of the material you have. If you had a larger amount of aluminum, for example, its density would still be about 3 g/cm^3. The density of water is always about 1 g/cm^3. It does not matter whether you have a single drop of water or a whole lake.

◀ Density is a measure of mass for each unit of volume. So the density of water is the same, about 1 g/cm^3, whether you have a tiny drop of dew or a whole lake.

▲ If an object's density is less than the density of water, that object will float in water.

▲ The liquids in this container all have different densities. Which liquid has the greatest density?

Whether an object sinks or floats in water depends on the object's density. An object will sink in water if it is made of a material with a density *greater than* the density of water. An object will float in water if it is made of a material with a density *less than* the density of water. An object's tendency to float is called buoyancy.

An ice cube floats in a glass of water because ice is less dense than water. Water has a density of about 1 g/cm^3, but ice has a density of about 0.9 g/cm^3. A piece of wood also floats in water because wood has a density of about 0.5 g/cm^3. Most rocks have a density of about 3 g/cm^3. This density is greater than that of water, so most rocks sink.

Liquids can float and sink, too. For example, if you pour cooking oil into water, the oil will settle in a layer above the water. This occurs because oil is less dense than water, so the oil actually floats on top of the water. If you pour corn syrup into water, the corn syrup sinks and forms a layer at the bottom of the container. You can probably guess why. Corn syrup is more dense than water.

 Explain how two blocks could have the same volume but different densities.

8

Solid

Solids have definite volumes and shapes. Fabric, stone, and plastic are only a few of the kinds of solid materials. Each kind of solid has different properties and is useful for different purposes.

particles in a solid

State of Matter

Most substances around us are in solid, liquid, or gas form. These forms are called **states of matter**. State of matter is a physical property.

All matter is made up of tiny particles that are constantly moving. (You will read more about these particles on pages 16 and 21.) A substance's state depends on the arrangement and movement of its particles.

A **solid** has a definite, or set, volume and a set shape because the particles in a solid are packed tightly together. They vibrate back and forth in place but do not move past one another. Stone is a solid. A block of stone will remain the same shape and take up the same amount of space, even if you move it from place to place.

Solid materials are used to make a variety of objects. Hard, strong solids such as wood, stone, and metal are used to make buildings. Soft, bendable solids such as fabric and rubber are used to make clothes and toys. Glass, paper, and plastic are other examples of solids. All these very different materials are the same in one way—they are all solids.

Liquid

The volume of milk in each container is the same, but the shape of the milk is different. This is because a liquid has no set shape but takes the shape of its container.

particles in a liquid

A **liquid** has a definite volume but no set shape because the particles in a liquid are not as tightly packed as in a solid. They can move around and slide past one another. Water, milk, and honey are examples of liquids.

Unlike solids, liquids have no set shape but instead take the shape of whatever container they are in. If you pour a liter of water from a bottle into a pail, the water will take the shape of the pail. However, the volume of the water will still be one liter. If the pail spills, the water will spread out onto the floor because there is no container to give it shape. However, the volume of water stays the same. You will still have to mop up one liter of water!

Because the particles in a liquid can move past one another, liquids flow easily from place to place. Juice, pancake syrup, and lava flowing from a volcano are other examples of liquids.

▲ Liquids pour and flow because the particles that make up liquids move easily past one another.

A **gas** has no set volume and no definite shape because the particles in a gas move freely and are far apart from one another. A gas will keep spreading out in all directions unless it is held in a container. Oxygen and helium are gases. The air we breathe is composed of many gases. If you blow up a balloon, the air takes the shape of the balloon. If you squeeze the balloon to change its shape and volume, the air inside changes its shape and volume, too.

Because many gases are invisible, it may seem that a gas does not have mass. However, recall that all matter has mass, and this is true even for a gas. Imagine you have two balloons of similar size and you fill one with air. The full balloon now has a greater mass because of the air it contains.

A fourth state of matter, called *plasma*, also exists. Plasma, the most common state of matter in outer space, is found in all stars, including the Sun. Plasma is similar to a gas but is super hot and glows. It is very rare on Earth but can be found in lightning, fluorescent lights, and some television sets.

 Describe the three main states of matter, and give an example of each one.

States of Matter

	Set Volume	Set Shape
Solid	yes	yes
Liquid	yes	no
Gas	no	no

▲ The different states of matter have volumes and shapes that may or may not change.

particles in a gas

Gas
As you pump air into a tire, the air fills whatever space is inside the tire. You could pump a little air into the tire, or a lot of air, because gas does not have a set volume.

▲ Lightning is in a super-hot, glowing state of matter called plasma. Plasma is also found in the Sun and other stars.

Other Physical Properties

Recall that physical properties are characteristics of matter that can be observed or measured. Hardness, magnetism, and conductivity are three additional physical properties that help us identify, describe, and classify matter.

Hardness is a physical property that describes how easy it is for a material to scratch or be scratched by another material. The property of hardness is often used to help identify minerals, the substances that make up rocks.

Another physical property is the ability to be attracted by a magnet. *Magnetic* substances can be attracted to a magnet or can become magnets. The metals iron, cobalt, and nickel are all magnetic. Most forms of steel are magnetic because steel contains iron.

The physical property of *conductivity* refers to how easily electricity or heat passes through a material. Most metals are good conductors of both electricity and heat. This is why electrical wiring and cooking pots and pans are often made of metal. Plastic and rubber are poor conductors of electricity. Electrical wires are often covered with plastic or rubber to prevent electricity from flowing to objects outside the wire.

 What property of most metals makes them good choices for electrical wires and cooking pots?

The physical property of hardness is used to identify minerals. Diamond, the hardest known mineral, can scratch all other minerals but cannot be scratched. ▶

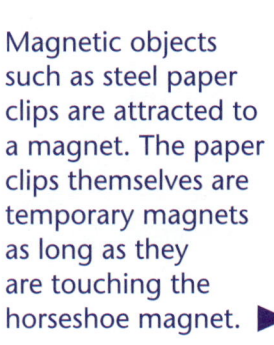

Magnetic objects such as steel paper clips are attracted to a magnet. The paper clips themselves are temporary magnets as long as they are touching the horseshoe magnet. ▶

▲ Most electrical cords have two parts. The wire inside is made of metal, which is a good conductor, and the coating outside is usually made of plastic, which is a poor conductor.

12

Chemical Properties of Matter

In addition to physical properties, matter also has chemical properties. **Chemical properties** describe the way one kind of matter, or substance, reacts with another.

Combustibility, or the *ability to burn*, is a chemical property. If you place wood in a fire, the wood will burn. If you place an iron nail in the same fire, its temperature will increase and it may even melt, but the iron will not burn. Substances that can burn are combustible. Wood is combustible but iron is not.

The *ability to rust* is another chemical property. Suppose you put the same piece of wood and the iron nail outside in the rain for a few days. When the wood dries, it will look about the same as it did before it got wet. The iron nail, however, will start to change color. It will get reddish brown spots of rust, a substance that forms when iron reacts with oxygen in the presence of water. Iron rusts but wood does not.

 Explain how chemical properties of matter differ from physical properties of matter.

A candle wick, natural gas, and wood are all combustible. Combustibility, the ability to burn, is a chemical property.

▲ The iron nail looks reddish brown because it has rusted. The ability to rust is a chemical property.

REFLECT ON READING

Before reading, you previewed the pictures in the book. What photograph or diagram in this section was helpful in illustrating new information about properties of matter? Explain to a partner the idea shown in the visual.

APPLY SCIENCE CONCEPTS

Choose a classroom object and write about it in your science notebook. Describe its properties. Include physical properties you can observe or measure and chemical properties such as its ability to burn or rust.

Build Reading Skills
Compare and Contrast

To **compare** objects or events is to notice how they are similar. To **contrast** objects or events is to notice how they are different.

As you read this section, try to identify similarities and differences between elements and compounds.

TIPS
Follow these guidelines for comparing and contrasting:
- Choose two related objects or events.
- To compare them, ask, "How are they alike? What is true about both of them?"
- To contrast them, ask, "How are they different? What is true about one that is not true about the other?"

A Venn diagram can help you organize similarities and differences.

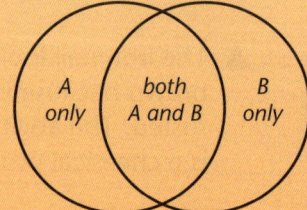

What Is Matter Made Of?

MAKE A CONNECTION

Did you ever open up an object or take it apart to discover what it was made of? This magnified view of the inside of a shell shows many details but not what the shell is made of. What do you think matter is made of?

FIND OUT ABOUT

- atoms, the tiny building blocks of matter
- elements and how they are grouped
- molecules
- compounds

VOCABULARY

atom, p. 16
nucleus, p. 16
proton, p. 16
neutron, p. 16
electron, p. 16
element, p. 17
periodic table, p. 18
metal, p. 19
molecule, p. 21
compound, p. 22

15

Here is additional information about some familiar elements listed in the periodic table. ▼

carbon

6 C Carbon

atomic number, 6
symbol, C
nonmetal
solid

Carbon is one of the most common nonmetal elements in nature. Graphite, the soft, slippery material used as pencil lead, is one form of the element carbon. Diamond, the hardest known mineral, is another.

silver

47 Ag Silver

atomic number, 47
symbol, Ag (from the Latin word for silver, *argentum*)
metal
solid

Silver is mined from Earth's crust and is used in jewelry, silverware, and coins. Because it is an excellent conductor, silver also is used in switches and other electrical devices.

neon

10 Ne Neon

atomic number, 10
symbol, Ne
nonmetal
gas

Neon is a gas element present in small amounts in the air. Its principal use is in advertising lighting because it glows when electric current runs through it.

iron

26 Fe Iron

atomic number, 26
symbol, Fe (from the Latin word for iron, *ferrum*)
metal
solid

Iron, the most common metal element inside Earth, is mined from iron ore in Earth's crust. Iron is usually mixed with carbon to make steel for buildings, tools, cars, rails, and other uses. Iron is highly magnetic and easily rusts in damp air.

Molecules

You know that all matter is made of atoms. The atoms in matter are often joined together in units called molecules. A **molecule** is a tiny particle of matter made of two or more atoms joined together.

For example, when two hydrogen atoms and one oxygen atom combine, they form a water molecule. A water molecule is the smallest unit of water that still has all the properties of water. If you could break a water molecule apart, you would no longer have the substance water.

When atoms of different elements combine, they form new substances with new properties.

- At room temperature, hydrogen is a gas. It is colorless, odorless, and lighter than air. It burns easily and can be used as rocket fuel.
- At room temperature, oxygen also is a colorless, odorless gas. It is part of the air we breathe, and it must be present in order for materials to burn.

The properties of these two elements are very different from the properties of water, which is a liquid at room temperature and which cannot burn.

Molecules may be composed of very few or very many atoms. Water molecules, for example, contain only three atoms. In contrast, the molecules of some plastics contain thousands of atoms.

 What are molecules?

A Water Molecule

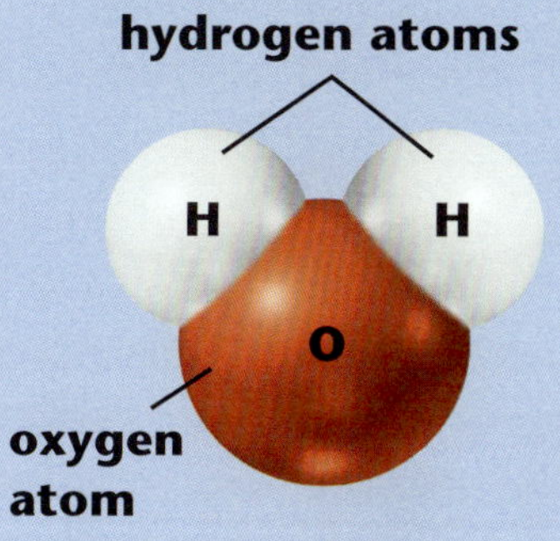

hydrogen atoms

H H

O

oxygen atom

Two hydrogen atoms and one oxygen atom combine to form one water molecule, the smallest unit of water.

▲ Water is the only substance that exists naturally on Earth in all three states of matter. The arrangement and movement of water molecules are different in solid ice, liquid water, and the gas water vapor.

Compounds

A **compound** is formed when two or more different elements combine chemically. Water is a compound composed of the elements hydrogen and oxygen. Table salt is a compound made of the elements sodium and chlorine. Atoms of these elements combine to form molecules of the compounds. A molecule is the smallest unit of a compound.

Most of the living and nonliving things around us are made of compounds. Fewer than 100 of the substances found naturally on Earth are elements. However, millions and millions of compounds on Earth are formed from this relatively small number of elements.

Scientists use a special way of writing called a chemical formula to show the makeup of a compound. A chemical formula includes chemical symbols and,

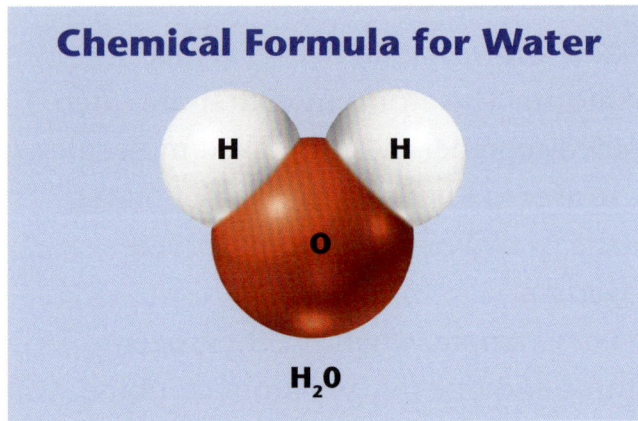

Chemical Formula for Water

▲ A chemical formula is a way to write the name of a compound. The chemical formula for water is H_2O.

sometimes, numbers. These symbols and numbers tell what elements, and how many atoms of each element, form a particle, or molecule, of the compound.

You know that each molecule of water contains two atoms of hydrogen and one atom of oxygen. Therefore, the chemical formula for water is H_2O. The chart on this page shows the chemical formulas of some familiar compounds.

The symbols in a chemical formula identify the elements in a compound. The numbers show how many atoms combine to form a molecule of the compound. Use the periodic table on pages 18–19 to identify the elements in these compounds. ▼

Common Compounds	
Compound	Chemical Formula
water	H_2O
carbon dioxide	CO_2
table salt	NaCl
glucose (a sugar)	$C_6H_{12}O_6$
table sugar	$C_{12}H_{22}O_{11}$
baking soda	$NaHCO_3$

table sugar

table salt

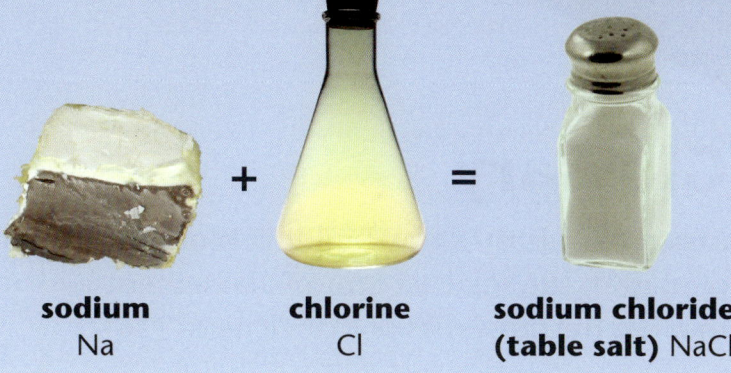

The elements that make up table salt are explosive sodium and poisonous chlorine. The properties of a compound are often unlike the properties of the elements that form it.

sodium
Na

chlorine
Cl

sodium chloride
(table salt) NaCl

The properties of compounds are often very different from the properties of the elements they are made of. Table salt, NaCl, is a good example. Sodium (Na) is a metal that explodes if it touches water, and chlorine (Cl) is a yellow-green poisonous gas. Together, they form a white crystal that is the very safe compound we use to flavor food.

Carbon combines with other elements to form the greatest number and variety of compounds. In fact, scientists have identified more than ten million different carbon compounds! Carbon dioxide, a gas in the air, is a compound of carbon and oxygen. Fossil fuels such as coal, oil, and natural gas are made of compounds of carbon and hydrogen. Carbon joins with calcium to form rocks such as limestone and marble. Most living things are made of compounds containing carbon.

Sugars are carbon compounds that living things use for energy. Green plants produce a kind of sugar, called glucose, which has the chemical formula $C_6H_{12}O_6$. Glucose provides energy for all living things—for plants themselves and for animals that eat plants.

 What is the main difference between compounds and elements?

Coal and other fossil fuels are made of compounds of carbon and hydrogen. ▶

▲ Green plants produce a very important compound called glucose, which provides energy for plants and for animals that eat plants.

REFLECT ON READING

Create a Venn diagram like the one on page 14, and use it to compare and contrast elements and compounds. Use the vocabulary words *atom* and *molecule* in your diagram.

APPLY SCIENCE CONCEPTS

Compare and contrast a gold ring and a cup of water. What type of substances are gold and water? How are they alike? How are they different? What is the basic unit of matter in each substance?

Glossary

atom (AT-uhm) a tiny building block of matter; the smallest unit of an element that has the properties of that element **(16)**

balance (BAL-uhns) a tool for measuring mass **(5)**

chemical property (KEM-i-kuhl PROP-ur-tee) a property of a substance that describes how the substance reacts with other substances **(13)**

compound (KOM-pound) a substance that is made up of the atoms of more than one element and that forms as the result of a chemical reaction **(22)**

density (DEN-si-tee) a measure of the amount of mass per unit volume of a substance **(7)**

electron (i-LEK-tron) a tiny part of an atom with a negative charge that moves in the space around the nucleus **(16)**

element (EL-uh-muhnt) a substance that cannot be broken down chemically into simpler substances and is made up of only one type of atom **(17)**

gas (GAS) the state of matter that has no definite volume and no definite shape **(11)**

liquid (LIK-wid) the state of matter that has a definite volume but not a definite shape **(10)**

mass (MAS) a measure of the amount of matter in an object **(5)**

matter (MAT-ur) anything that takes up space and has mass **(4)**

metal (MET-l) an element that is a solid, is usually shiny, and is a good conductor of heat and electricity **(19)**

molecule (MOL-i-kyool) the smallest unit of a substance that can exist alone and still have the properties of that substance **(21)**

neutron (NOO-tron) a tiny part of an atom with no charge that is located in the nucleus of the atom **(16)**

nucleus (NOO-klee-uhs) the central part of an atom, made up of protons and neutrons **(16)**

periodic table (pir-ee-OD-ik TAY-buhl) a chart in which elements are arranged according to their chemical properties **(18)**

physical property (FIZ-i-kuhl PROP-ur-tee) a characteristic of a substance that can be observed or measured **(4)**

proton (PROH-ton) a tiny part of an atom with a positive charge that is located in the nucleus of the atom **(16)**

solid (SOL-id) the state of matter that has a definite volume and a definite shape **(9)**

state of matter (STAYT uhv MAT-ur) the physical form of a substance **(9)**

volume (VOL-yoom) a measure of the amount of space an object takes up **(6)**